ICS 93.160
P 59
File No: J 838—2009

Standard of Ministry of Water Resources of the People's Republic of China

SL 142—2008
Replace SL 142—97

Code of Practice for Model Acceptance Tests of Hydraulic Turbine with Sediment Water

Issued on November 10, 2008   Implemented on February 10, 2009

Yellow River Conservancy Press
· Zhengzhou ·

图书在版编目(CIP)数据

水轮机模型浑水验收试验规程＝Code of practice for model acceptance tests of hydraulic turbine with sediment water：英文/中华人民共和国水利部发布. —郑州：黄河水利出版社，2013.12
ISBN 978-7-5509-0535-1

Ⅰ.①水… Ⅱ.①中… Ⅲ.①水轮机-模型-验收试验-规程-英文 Ⅳ.①TK73-65

中国版本图书馆 CIP 数据核字(2013)第 203728 号

出 版 社：黄河水利出版社
　　　　　　地址：河南省郑州市顺河路黄委会综合楼 14 层　邮政编码：450003
发行单位：黄河水利出版社
　　　　　　发行部电话：0371-66026940、66020550、66028024、66022620(传真)
　　　　　　E-mail：hhslcbs@126.com
承印单位：黄河水利委员会印刷厂
开本：850 mm×1 168 mm　1/32
印张：1.375
字数：65 千字　　　　　　　　　　　　印数：1—1 000
版次：2013 年 12 月第 1 版　　　　　　印次：2013 年 12 月第 1 次印刷

定价：80.00 元

Introduction of English Version

Department of International Cooperation, Science and Technology of Ministry of Water Resources, P. R. China has the mandate of managing the formulation and revision of water technology standards in China.

Translation of this English version of standard was organized by Department of International Cooperation, Science and Technology of Ministry of Water Resources, P. R. China in accordance with due procedures and regulations applicable in the country.

This English version of standard is identical to its Chinese original *Code of Practice for Model Acceptance Tests of Hydraulic Turbine with Sediment Water* (SL 142—2008), which was formulated and revised under the auspices of Department of International Cooperation, Science and Technology of Ministry of Water Resources, P. R. China.

Translation of this standard was undertaken by China Institute of Water Resources and Hydropower Research, Tsinghua University and China Agricultural University.

Translation team includes Lu Li, Liao Cuilin, Xu Hongyuan, Wang Fujun, Ma Suping, Luo Xianwu, Tang Xuelin and Wang Wanpeng.

This standard was reviewed by Liu Zhiming and Wang Fujun.

Department of International Cooperation, Science and Technology
Ministry of Water Resources, P. R. China

Foreword

Code of Practice for Model Acceptance Tests of Hydraulic Turbine with Sediment Water (SL 142—97) is revised based on the requirements of *Specification for the Drafting of Technical Standards of Water Resources* (SL 1—2002). This revision is according to the requirements of revising *The Investigation, Design and Research of Water Resources and Hydropower Technical Standards* required by General Institute of Water Resources and Hydropower Planning and Design, Ministry of Water Resources.

This code contains 7 chapters, 15 sections, 108 articles and 2 annexes. The main technical contents are as follows:

—General provisions.
—Terms, symbols and units.
—Test rig.
—Model turbine.
—Parameter measurement and uncertainty.
—Acceptance tests.
—Verification of guarantees.

The revised contents of SL 142—97 mainly include the following:

—Foreword and basic information are added to the introduction part and the additional explanations in the original code are deleted.

—In the chapter "Terms, symbols and units", the English translations for terms are added and some terms are amended.

—In the chapter "Test rig", the required minimum test Reynolds number is increased and the required uncertainty of efficiency testing for clean water and sediment water is improved.

—In the chapter "Model turbine", the requirements for manufacture deviation of hydraulic passage of model turbine are improved.

—In the chapter "Parameter measurement and uncertainty", the requirements for median grain size of sediment for the test are added.

—In the chapter "Verification of guarantees", evaluation of relative erosion resistant capacity is added.

—Chapter 8 "Management of model acceptance tests" in the original code is deleted and Annex B "Acceptance test program and report" is added.

In addition, the local structure and word description in the original code are amended.

This code is to replace the original edition SL 142—97.

This code was approved by Ministry of Water Resources of the People's Republic of China.

This code was initiated by General Institute of Water Resources and Hydropower Planning and Design, Ministry of Water Resources.

This code will be explained by General Institute of Water Resources and Hydropower Planning and Design, Ministry of Water Resources.

This code was chiefly drafted by China Institute of Water Resources and Hydropower Research.

This code was jointly drafted by China Water Resources Beifang Investigation, Design and Research Co. Ltd.

This code is published and distributed by China Water & Power Press.

Chief drafters of this code are Lu Li, Li Tieyou, He Chenglian, Wang Haian and Ma Suping.

The technical responsible persons of this code review conference are Liu Zhiming and Wang Fujun.

The format examiner of this code is Dou Yisong.

Contents

1 **General Provisions** ·· (1)
2 **Terms, Symbols and Units** ···························· (2)
   2.1 Terms ·· (2)
   2.2 Symbols and Units ······························· (7)
3 **Test Rig** ·· (9)
4 **Model Turbine** ·· (13)
5 **Parameter Measurement and Uncertainty** ·············· (21)
   5.1 General Requirements ···························· (21)
   5.2 Discharge Measurement and Uncertainty ············ (21)
   5.3 Head Measurement and Uncertainty ·················· (21)
   5.4 Shaft Torque Measurement and Uncertainty ········ (22)
   5.5 Rotational Speed Measurement and Uncertainty ··· (22)
   5.6 Sediment Concentration Measurement and Deviation ·· (23)
   5.7 Other Parameters Measurement and Uncertainty ··· (23)
   5.8 Cavitation Coefficient and Efficiency of Model Turbine ··· (24)
   5.9 Total Uncertainty ································· (24)
6 **Acceptance Tests** ·· (26)
   6.1 Efficiency Test ···································· (26)
   6.2 Cavitation Test ···································· (26)
   6.3 Sand Erosion Test ································· (28)
   6.4 Other Tests ··· (28)
7 **Verification of Guarantees** ···························· (30)

Annex A Density of Mercury, Physical Property of
         Water and Acceleration Due to Gravity ········ (32)
Annex B Acceptance Test Program and Report ············ (34)
Explanation of Code Terminology ····························· (36)

1 General Provisions

1.0.1 This code is formulated for standardizing model acceptance tests of hydraulic turbine with sediment water.

1.0.2 This code is applicable to model acceptance tests of reaction turbines (including Francis, diagonal and axial turbines) with sediment water in which the suspended-sediment concentration does not exceed 50 kg/m^3 and median grain size is less than 0.1 mm.

1.0.3 Model acceptance tests of hydraulic turbine with sediment water consist of efficiency, cavitation, sand erosion and other tests.

1.0.4 The following standards contain provisions which, through reference in this text, constitute provisions of this code.

Electrotechnical Terminology-hydroelectric Powerplant Machinery (GB/T 2900.45)

Specification for Water Passage Components of Hydraulic Turbines (GB/T 10969)

Code for Model Acceptance Tests of Hydraulic Turbines (GB/T 15613)

Hydraulic Turbines, Storage Pumps and Pump-turbines-model Acceptance Tests (IEC 60193)

1.0.5 Not only the requirements stipulated in this code, but also those in the current relevant ones of the nation shall be complied with during model acceptance tests of hydraulic turbine with sediment water.

2 Terms, Symbols and Units

2.1 Terms

2.1.1 clean water

Suspended-sediment concentration is not larger than 0.05 kg/m^3 and cavitation bubbles at the runner inlet or outlet can be clearly observed visually.

2.1.2 sediment water

Suspended-sediment concentration is larger than 0.05 kg/m^3 and cavitation bubbles at the runner inlet or outlet can not be clearly observed visually.

2.1.3 sediment concentration

Mass of sediment per unit volume of water.

2.1.4 water column

Height of clean water column used to express pressure.

2.1.5 density of clean water

Mass per unit volume of clean water.

2.1.6 density of sediment water

Mass per unit volume of sediment water.

2.1.7 gauge pressure

Reading data on the gauge larger or less than atmospheric pressure at any measuring point in the system.

2.1.8 gauge height

Height a of the gauge location in elevation above the measuring point is converted to water column and calculated according to Equation (2.1.8):

$$a' = \frac{\rho_s a}{\rho} \qquad (2.1.8)$$

2.1.9 potential head

Height Z from the measuring point to the reference plane is converted to water column with reference to density of sediment water and calculated according to Equation (2.1.9):

$$Z' = \frac{\rho_s Z}{\rho} \tag{2.1.9}$$

2.1.10 pressure head

Pressure at a point in the system is expressed as water column and calculated according to Equation (2.1.10):

$$h_p = a' + \frac{p_g}{\rho g} \tag{2.1.10}$$

2.1.11 velocity head

Velocity head is a value of square of mean velocity divided by 2 times acceleration due to gravity. It is expressed by water column in reference to the density of sediment water and calculated according to Equation (2.1.11):

$$h_v = \frac{\rho_s V^2}{2\rho g} \tag{2.1.11}$$

2.1.12 head

The head is the sum of the pressure head, velocity head, and potential head at a given section and is calculated according to Equation (2.1.12):

$$H_t = h_p + h_v + Z' \tag{2.1.12}$$

2.1.13 net head

Net head is the effective head available to the turbine unit for power production (see Figure 2.1.13).

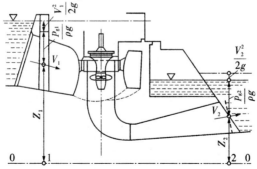

a) Reaction turbine with vertical shaft, spiral case with trapezoidal section and elbow type draft tube

$$H_n = \left[(Z_1 - Z_2) + \frac{V_1^2 - V_2^2}{2g} \right] \frac{\rho_s}{\rho} + \frac{p_{g1} - p_{g2}}{\rho g}$$

Figure 2.1.13 Determination for net head of turbine

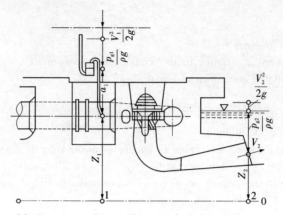

b) Reaction turbine with vertical shaft, spiral case with round section and elbow type draft tube

$$H_n = \left[(Z_1 + a_1 - Z_2) + \frac{V_1^2 - V_2^2}{2g} \right] \frac{\rho_s}{\rho} + \frac{p_{g1} - p_{g2}}{\rho g}$$

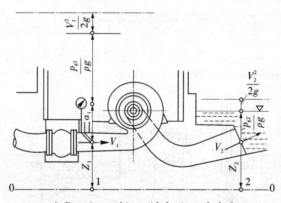

c) Reaction turbine with horizontal shaft

$$H_n = \left[(Z_1 + a_1 - Z_2) + \frac{V_1^2 - V_2^2}{2g} \right] \frac{\rho_s}{\rho} + \frac{p_{g1} - p_{g2}}{\rho g}$$

Continued to Figure 2.1.13

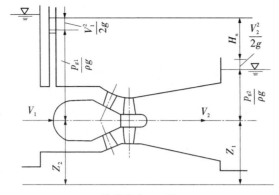

d) Tubular turbine

$$H_n = \left[(Z_1 - Z_2) + \frac{V_1^2 - V_2^2}{2g}\right]\frac{\rho_s}{\rho} + \frac{p_{g1} - p_{g2}}{\rho g}$$

Continued to Figure 2.1.13

2.1.14 static suction head

Static suction head is the value of the elevation of the reference plane of hydraulic turbine minus the tailwater level (see Figure 2.1.14). It is converted to water column in reference to the density of sediment water and calculated according to Equation (2.1.14):

$$H'_s = \frac{\rho_s H_s}{\rho} \qquad (2.1.14)$$

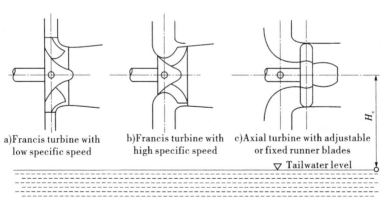

a) Francis turbine with low specific speed
b) Francis turbine with high specific speed
c) Axial turbine with adjustable or fixed runner blades

Figure 2.1.14 The reference plane for determination of static suction head H'_s

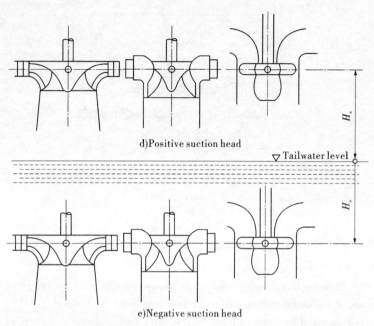

Continued to Figure 2.1.14

2.1.15 reference runner diameter

Reference runner diameter is the runner diameter of the given part (see Figure 2.1.15).

2.1.16 grain size distribution

Sediment is classified by grain size. Grain size distribution is defined by a ratio of the weight of one class of grain size to the total weight.

2.1.17 median grain size

The grain size corresponds to the point at 50% (by weight) of the grain-size distribution curve.

2.1.18 sediment mineral composition

Sediment mineral composition accounts for the content of different mineralogical compositions in sediment.

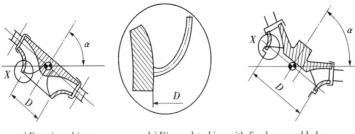

a) Francis turbine   b) Diagonal turbine with fixed runner blades and runner band

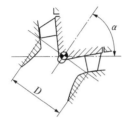 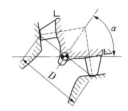

c) Diagonal turbine with fixed runner blades, without runner band   d) Diagonal turbine with adjustable runner blades

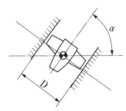 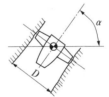

e) Axial turbine with fixed runner blades   f) Axial turbine with adjustable runner blades

Figure 2.1.15　Reference runner diameter

2.1.19　sand erosion

Sand erosion is the damage to the surface material of hydraulic passage parts caused by sediment water flow.

2.2　Symbols and Units

S—sediment concentration, kg/m^3;

h — water column, m;
ρ — density of clean water, kg/m³;
ρ_s — density of sediment water, kg/m³;
Z' — potential head, m;
h_p — pressure head, m;
a' — gauge height, m;
p_g — gauge pressure, Pa;
h_v — velocity head, m;
H_t — head, m;
H_n — net head, m;
h_a — atmospheric pressure, m;
H'_s — static suction head, m;
σ — cavitation coefficient;
σ_c — critical cavitation coefficient;
σ_i — incipient cavitation coefficient;
σ_p — plant cavitation coefficient;
Q — turbine discharge, m³/s;
V — mean velocity, m/s;
P_{in} — turbine input power, kW;
P_{out} — turbine output power, kW;
η — turbine efficiency;
n — rotational speed, r/min;
n_{run} — runaway speed, r/min;
d_{50} — median grain diameter, mm.

3 Test Rig

3.0.1 There are two types of test rigs, closed-loop and open test rigs. Generally open test rig is not applied to cavitation test of model turbine with sediment water.

3.0.2 Stirring devices for sediment water shall be installed at the upstream and downstream sides of open test rig. The shape of water tanks at upstream and downstream sides shall be designed in such way that could prevent the sediment deposition for the closed test rig. If necessary, stirring devices of sediment water shall be set.

3.0.3 The sediment-adding device in the closed test rig should be installed at the top of downstream water tank. The sediment-adding velocity shall be controlled and the sealing performance shall be kept after the completion of adding sediment.

3.0.4 The pipeline diameter in test rig shall meet the requirements for sediment starting velocity.

3.0.5 The quick discharge valve shall be installed at the bottom of the test rig system.

3.0.6 The model tests with clean water should be carried out at the same test rig before model acceptance tests (including efficiency test, cavitation test and other tests) with sediment water.

3.0.7 Efficiency test, cavitation test and other tests of hydraulic turbine may be performed under different test heads and the determination of test heads shall be in accordance with the following requirements.

    a For efficiency test, test head shall not be less than 4 m for axial turbine, and 20 m for Francis and diagonal turbines.

    b For cavitation test, test head shall not be less than 8 m (the test head may be reduced suitably at large discharge operating condition out of optimum operating condition) for Kaplan and propeller turbines, 4 m for tubular turbine and 20 m for Francis and diagonal turbines. If the head for prototype Kaplan and propeller turbines is less than 8 m, the test head may be reduced suitably but shall not be less than 4 m.

    c Test head for sand erosion test shall not be less than that for efficiency test.

    d Test head for hydraulic pressure fluctuation test shall not be less than that for cavitation test.

3.0.8 The minimum model dimension and the test condition shall meet the requirements in Table 3.0.8.

Table 3.0.8 **The requirements of the minimum model dimension and the test condition**

Type	Axial and diagonal turbines	Francis turbine
Re	$>4\times10^6$	
D(mm)	$\geqslant 300$	$\geqslant 250$

Reynolds number Re shall be calculated according to Equation (3.0.8):

$$Re = \frac{Du}{\nu} \quad (3.0.8)$$

where D—reference runner diameter, m;
          u—peripheral velocity at the reference runner diameter D, m/s;
          ν—kinematic viscosity, m²/s.

3.0.9 The test head shall be steady. The permissible differential between maximum and minimum heads shall be within 0.5%❶ of test head in the process of measuring discharge, rotational speed and torque.

3.0.10 Upstream measuring point for measuring test head shall be arranged at the inlet of spiral case and the downstream measuring point shall be arranged at the outlet of draft tube [see Figure 2.1.13a), b), c)]. Velocity head may be calculated by using the mean velocity of the measuring section.

3.0.11 If static pressure measurement is adopted, at least 4 static holes shall be set on each measuring section. If the section is round, 4

---

    ❶ In the Chinese version code, it is ±0.5%. In order to express correctly, it is amended to 0.5% in this English version code.

static holes shall be arranged at mutually perpendicular diameters; if the section is rectangle, 4 static holes shall be set at the midpoint of each edge. Each static hole shall be connected with a pressure gauge by a pressure-balancing loop pipe or pipes having the same flow resistance. If the measuring section is set at the horizontal pipe, static holes shall not be arranged at the highest point and lowest point of the section. Besides, an exhaust valve and a sand flushing valve shall be arranged at the highest point and lowest point respectively (see Figure 3.0.11). If the measuring section is set at a vertical pipe, at least 4 exhaust and sand flushing valves shall be arranged at the pressure-balancing loop pipe.

3.0.12 For static pressure measuring procedures, the static hole shall be circular and perpendicular to the inner surface of the pipe. The measuring hole shall be 3-6 mm in diameter and at least twice the diameter long, and its fillet shall not be larger than $d/4$. The inner surface of pipe shall be smooth within a range of at least 100 mm away from the measuring point (see Figure 3.0.12).

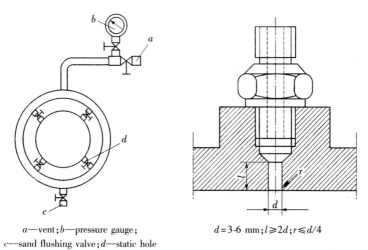

a—vent; b—pressure gauge;
c—sand flushing valve; d—static hole

Figure 3.0.11 Arrangement of static holes and measuring section

$d = 3\text{-}6$ mm; $l \geqslant 2d$; $r \leqslant d/4$

Figure 3.0.12 Form and dimension of static hole

3.0.13 The submerged water level at the inlet of the model turbine shall be at least 1 m higher than that at the highest point of the inlet section and the submerged water level at the outlet shall be at least 0.5 m higher than that at the highest point of the draft tube outlet section.

3.0.14 The inlet flow of the model turbine shall be irrotational and velocity shall be uniform.

3.0.15 No water shall flow into or out of the system between the model turbine and the discharge measuring device.

3.0.16 Sand used for model acceptance tests should be the same as that passing through the prototype turbine.

3.0.17 When air bubbles are released during a cavitation test, it shall be guaranteed that normal works of the measuring instruments for the test head and discharge are not influenced. When water in the closed test rig is replaced or compensated by new water, the formal test shall be performed after the test rig works more than one hour under vacuum degree of 7-8 m of the water column. When sediment in the closed test rig is replaced or compensated by new sediment, the formal test shall be performed after the test rig works more than 20 min at the same vacuum degree.

3.0.18 The maximum vacuum degree inside the tailwater tank should not exceed 8 m water column during a cavitation test.

3.0.19 The test rig for model acceptance tests shall be technically verified and the primary measuring equipments shall have calibration certificates issued by authorized measurement inspection institutions.

For the accuracy of the test rig, the permissible total uncertainty of efficiency shall be within ±0.3% for clean water and within ±0.5% for sediment water.

3.0.20 The actual uncertainty of the test rig shall refer to calibration results for model acceptance tests.

4 Model Turbine

4.0.1 Hydraulic passage parts of model turbine shall meet sand erosion requirements during the test with sediment water. The efficiency variation due to the change of flow passage surface caused by sand erosion shall be within the total uncertainty of the test rig after the test with sediment water.

4.0.2 Hydraulic passage parts of model turbine (including spiral case, stay ring, guide vane, bottom ring, runner chamber and draft tube, etc.) shall be manufactured strictly according to design drawings. The permissible deviation of geometrical shape according to the design drawings shall be in accordance with those specified in Table 4.0.2-1. The surface roughness of the parts shall be in accordance with those specified in Table 4.0.2-2.

Table 4.0.2-1 Permissible geometrical deviations for hydraulic passage parts of model turbine (see Figure 4.0.2-1 and Figure 4.0.2-2)

Component	Item	Permissible dimensional deviation (%)	Remarks
Spiral case	Inlet diameter D_s	±1.0	
	Dimension A, B, C, E	±1.5	
	Center distance R	±1.0	
Stay ring	Inner diameter D_a, outer diameter D_b	±0.4	
	Arc radius R	±2.0	
	Stay vane height B_0	±0.3	
	Stay vane profile	±3.0	The ratio of the thickness deviation to the maximum thickness of stay vane

Continued to Table 4.0.2-1

Component	Item	Permissible dimensional deviation (%)	Remarks
Guide vane	Guide vane height b_0	±0.2	
	Guide vane pitch circle diameter D_0	±0.1	
	L_1, L	±1.0	
	Guide vane profile	±2.0	The ratio of the thickness deviation to maximum thickness of guide vane
	Waviness	±0.02	
Blade	Blade profile	±0.1D_2	
	Waviness	±0.02	
Bottom ring	Inner diameter D_d	±0.1	
	Arc radius R	±2.0	
Runner chamber	Throat diameter D_{th}	±0.2	
	Diameter of runner chamber D_1	±0.1	
Draft tube	Inlet diameter D_1	±1.0	
	$L_1, H_1, H_2, W_1, W_2, W$	±2.0	

Table 4.0.2-2 Permissible maximum roughness of hydraulic passage surface of model turbine

Component	Roughness R_a (μm)
Runner and guide vane	0.8 ~ 3.2
Stay ring, stay vane, bottom ring, runner chamber, runner cone, head cover, etc.	3.2 ~ 12.5

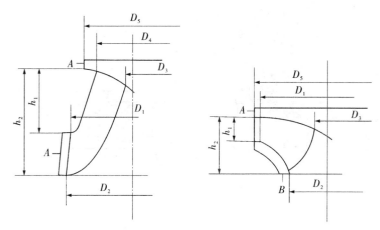

Figure 4.0.2-1　Principal dimensions of Francis turbine runner

4.0.3 Assembly dimensions and clearance of model turbine shall be in accordance with those specified in Table 4.0.3. Shaft seals shall be appropriate to work in sediment water and no leakage of sediment water occur. Friction force shall be steady during a test. If the shaft seal is lubricated by external water, the lubricating-water quantity which flows into the flow passage of the model turbine shall be deducted from the total discharge through the turbine.

4.0.4 Some parts of model turbine may be made of transparent and nonmetallic material, but the surface roughness, dimension tolerance and the stiffness of the parts shall be guaranteed.

4.0.5 For model turbine with adjustable runner blades, the centerline of the blade axis shall not be moved and the blade tip clearance shall not be adjusted when the blades are rotated. Permissible deviation of a single blade angle shall be within $\pm 0.25°$.

4.0.6 Surfaces shall be kept smooth and flat at the joints of hydraulic passage parts of model turbine.

4.0.7 The following items shall be checked: principal dimensions of the spiral case, stay vane, guide vane, runner, runner chamber and draft tube; the number of stay vane, guide vane and runner blades; clearances of the runner and guide vanes.

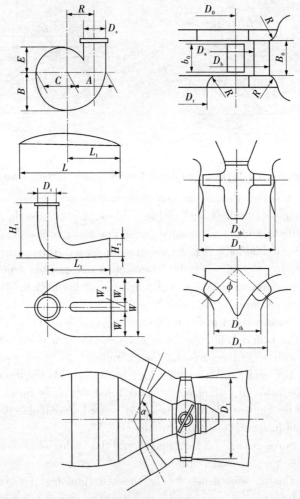

Figure 4.0.2-2 Principal dimensions of axial and diagonal turbines

Table 4.0.3 Requirements for assembly dimensions of model turbine
(see Figure **4.0.3**)

Item	Permissible value(%)	Remarks
a	$0.05<a<0.1$	Ratio of a to diameter at the corresponding point
b	>0.1	Ratio of b to guide vane height
c	$0.05<c<0.08$	Ratio of c to diameter of runner chamber
d	≈ 0.5	Ratio of d to diameter of runner hub
e	$0.035<e<0.08$	Ratio of e to reference runner diameter
a_0	± 2.0	Ratio of a_0 to optimum average opening

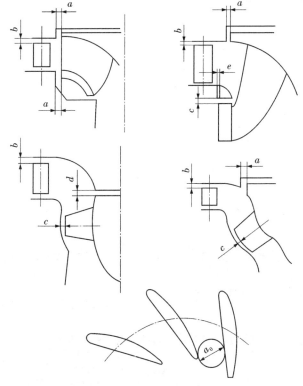

Figure 4.0.3 Check position for assembly dimensions of model turbine

4.0.8 The check for profile and dimensions of runners shall be in accordance with the following requirements.

  a For Francis turbine, the shape and opening of the blade inlet and outlet shall be checked at least at the top, middle and low sections using a surveying instrument or template (see Figure 4.0.8-1 and Figure 4.0.8-2). The principal dimension deviations of the runner, inlet and outlet blade angles on the middle section (see Table 4.0.8-1), and runner radial and end-face runout shall be checked (see Table 4.0.8-2).

  b For axial and diagonal turbines, the blade profile shall be checked on at least 4 sections using surveying instrument or template, and at least 5 measuring points shall be arranged on each section. The principal dimension deviations of runner shall be checked (see Table 4.0.8-3).

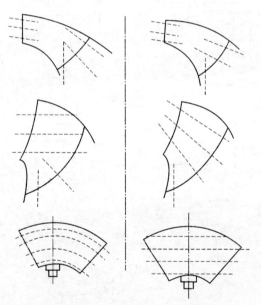

Figure 4.0.8-1 Check sections of blade

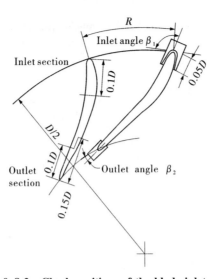

Figure 4.0.8-2 Check positions of the blade inlet and outlet

Table 4.0.8-1 Check positions and permissible deviations for model runner of Francis turbine

Item	Permissible deviation		Remarks	
	Average value	Individual value		
D_1, D_3	±0.20%	±0.40%	The ratio of the deviation to corresponding diameter shall be checked according to manufacturing drawing	See Figure 4.0.2-1, Figure 4.0.2-2 and Figure 4.0.8-2
D_2, D_4	±0.25%	±0.35%		
D_5	±0.50%	±0.75%		
h_1, h_2	±0.50%	±0.75%		
β_1	±1.5°	±3°		
β_2	±1°	±2°		

4.0.9 Template or other special measuring instruments shall be provided by the model turbine manufacturer. The template shall be checked according to permissible deviation and the deviation of the template shall not be larger than the permissible deviation of the blade profile.

Table 4.0.8-2 Check positions and permissible values of radial runout and end-face runout for model runner of Francis turbine

Item	Permissible value (%)	Remarks
A	0.01	The ratio of the deviation to corresponding diameter shall be checked according to Figure 4.0.2-1
B	0.01	

Table 4.0.8-3 Check positions and permissible values of radial runout and end-face runout of model runners of axial and diagonal turbines

Item	Permissible value	Remarks
D_1	It is determined according to dimensions of runner chamber and requirements set out in Table 4.0.8-1	See Figure 4.0.8-3
D_b, D_3, D_4	±0.1%	
D_5	Check according to manufacturing drawing	
ϕ	±0.25°	

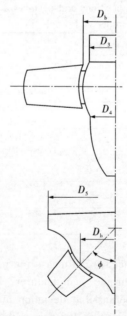

Figure 4.0.8-3 Main check positions of the model runners of axial and diagonal turbines

5 Parameter Measurement and Uncertainty

5.1 General Requirements

5.1.1 The main measuring instruments shall be calibrated before and after a model acceptance test.

5.1.2 Measuring instruments for head, shaft torque, etc. shall be calibrated before and after each test.

5.2 Discharge Measurement and Uncertainty

5.2.1 Discharge measurement may adopt primary method (including weighing method and volumetric method), secondary method (electromagnetic flowmeter), etc. Sediment in tank shall be cleaned out before measuring when using volumetric method.

5.2.2 It takes at least 50 s to fill a weighing or volumetric tank with water for weighing method or volumetric method. The switch device for split flow and filling-water time shall be accurate.

5.2.3 Flow meter shall be calibrated with clean water by primary method when electromagnetic flow meter is used to measure discharge. If possible, flow meter shall be calibrated with sediment water.

5.2.4 Permissible uncertainty of discharge measurement shall be in accordance with those specified in Table 5.2.4.

Table 5.2.4 Permissible uncertainty of discharge measurement

Measurement method	Clean water(%)	Sediment water(%)
Weighing method	±0.1	±0.2
Volumetric method	±0.1	±0.2
Electromagnetic flow meter	±0.2	±0.3

5.3 Head Measurement and Uncertainty

5.3.1 The head may be measured by a pressure transducer, differential pressure transducer or other high-accuracy pressure gauge.

5.3.2 Permissible uncertainty of the head measurement shall be within ±0.15%.

5.4 Shaft Torque Measurement and Uncertainty

5.4.1 Torque and rotational speed of the turbine shaft shall be measured by using a torque measurement device, and then the output power of model turbine shall be calculated according to Equation (5.4.1).

$$P_{out} = M\omega \qquad (5.4.1)$$

where P_{out} —the output power of main shaft of model turbine, W;

M —the torque of main shaft of model turbine, N · m;

ω —the angular velocity of main shaft of model turbine, rad/s.

5.4.2 The torque measurement device (such as dynamometer motor, torque meter, etc.) shall have good stability and high sensitivity.

5.4.3 The torque measurement arm shall be made of metallic material insensitive to external temperature. The permissible length error of the torque arm shall be within ±0.02%.

5.4.4 The balance force to calculate the torque shall be measured using national standard weights or the calibrated electrical-load transducer. The permissible uncertainty of balance force measurement shall be within ±0.12% for clean water.

5.5 Rotational Speed Measurement and Uncertainty

5.5.1 The primary signal device for rotational speed measurement shall be rigidly connected with the main shaft of model turbine to ensure no relative movement.

5.5.2 The rotational speed measurement may be done with an electromagnetic induction device, photoelectric pulse generator or other instruments that measure rotational speed.

5.5.3 The permissible fluctuation value of rotational speed shall be within ±0.25% in the process of measuring the rotational speed.

5.5.4 When measuring runaway speed, the main shaft and the torque measurement device shall be disconnected, or the torque load on the shaft shall be adjusted to zero under the connection condition.

5.5.5 The permissible uncertainty of rotational speed measurement shall be within ±0.1%.

5.6 Sediment Concentration Measurement and Deviation

5.6.1 The measuring position of sediment concentration shall be arranged at the penstock in front of the spiral case inlet. No water flow into or out of the system between the measuring point and model turbine, and the penstock shall be straight with constant diameter. Measurement of sediment concentration shall not affect the measuring results of pressure at the turbine inlet.

5.6.2 The median grain size of sediment shall be checked and analyzed before test. The permissible deviation of median grain size measured at the beginning and end of the test shall be within ±15%.

5.6.3 Weighing method combined with sampling by filtration or other methods may be adopted to measure the sediment concentration. The sampling point shall be arranged in the center of the penstock when using the sampling filtering weighing method. When secondary-class method is adopted in a test, it shall be calibrated using the weighing method combined with the sampling by filtration, and the permissible difference between two methods shall be within ±10%.

5.6.4 The permissible deviation between the sediment concentration prescribed in test project and that measured at the design load shall be within ±10%.

5.6.5 If a bypass pipe is used to measure the sediment concentration, the discharge through the bypass pipe shall be deducted from the total discharge through the turbine.

5.6.6 Sediment concentration shall be measured at least twice under the same operating conditions at the beginning and end of each test. Permissible deviation between them shall be within ±5%.

5.7 Other Parameters Measurement and Uncertainty

5.7.1 Measurement of static suction head: permissible measuring uncertainty shall be within ±0.2%.

5.7.2 Measurement of atmosphere pressure: permissible measuring uncertainty shall be within ±0.2%.

5.7.3 Measurement of vacuum degree: permissible measuring uncertainty shall be within ±0.2%.

5.7.4 Measurement of water temperature: permissible measuring uncertainty shall be within ±0.5 ℃.

5.7.5 Measurement of hydraulic pressure fluctuation: permissible measuring uncertainty of the pressure fluctuation transducer shall be within ±0.5%.

5.8 Cavitation Coefficient and Efficiency of Model Turbine

5.8.1 Cavitation coefficient σ of turbines shall be calculated according to Equation (5.8.1):

$$\sigma = \frac{h_a - h_{va} - h_{vt} - H'_s}{H_n} \qquad (5.8.1)$$

where　σ—cavitation coefficient of the turbine;

h_a—atmosphere pressure, m;

h_{va}—vapour pressure of water, m;

h_{vt}—vacuum degree at water level at draft tube outlet, m;

H'_s—static suction head, m;

H_n—net head, m.

5.8.2 Efficiency of model turbine shall be calculated based on the measured parameters (such as head, discharge, torque and rotational speed) according to Equation (5.8.2):

$$\eta_M = \frac{P_{out,M}}{P_{in,M}} = \frac{M\omega}{\rho g Q H_n} = \frac{P'L\pi n}{30\rho g Q H_n} = K\frac{P'Ln}{\rho g Q H_n} \qquad (5.8.2)$$

where　P'—balance force of torque measurement device, N;

L—arm length of torque measurement device, m;

n—rotational speed of turbine, r/min;

H_n—net head in test and expressed as water column, m;

Q—turbine discharge, m^3/s;

ρ—density of clean water, kg/m^3.

5.9 Total Uncertainty

5.9.1 Total uncertainty shall include systematic uncertainty and random uncertainty. The systematic uncertainty is subjected to measurement methods and the calibration results of measuring instruments. Random uncertainty shall be determined according to the principle of statistics and the number of measuring times shall be enough to reduce random uncertainty to a minimum value.

Systematic uncertainty of the model efficiency shall be calculated according to Equation (5.9.1-1):

$$E_{\eta s} = \pm \sqrt{E_Q^2 + E_{H_n}^2 + E_{P'}^2 + E_n^2 + E_L^2} \qquad (5.9.1\text{-}1)$$

where E_Q—the systematic uncertainty of discharge measurement, %;

E_{H_n}—the systematic uncertainty of net head measurement, %;

$E_{P'}$—the systematic uncertainty of balance force measurement, %;

E_n—the systematic uncertainty of rotational speed measurement, %;

E_L—the systematic uncertainty of force arm measurement, %.

Permissible random uncertainty $E_{\eta R}$ of model efficiency shall be within ±0.15% for clean water and ±0.25% for sediment water.

The total uncertainty of efficiency of model turbine is the combination of the systematic and random uncertainty by the root-sum-square method and calculated according to Equation (5.9.1-2):

$$E_\eta = \pm \sqrt{E_{\eta s}^2 + E_{\eta R}^2} \qquad (5.9.1\text{-}2)$$

5.9.2 Density of mercury, vapour pressure of water, density of clean water, kinematic viscosity of clean water at different temperatures, acceleration due to gravity at different latitudes and elevations are shown in Annex A. Vapour pressure and dynamic viscosity of sediment water at different temperatures is taken as that of clean water temporarily.

6 Acceptance Tests

6.1 Efficiency Test

6.1.1 Efficiency test should be performed with plant cavitation coefficient. If efficiency test is performed under no cavitation condition, whether the efficiency is affected by cavitation shall be checked through the test under the condition of plant cavitation coefficient. If the efficiency is affected, the efficiency shall be corrected according to the test results.

6.1.2 Efficiency test with clean water shall be performed firstly before efficiency test with sediment water. Efficiency and power tests with clean water should be performed at 30%-120% of guide vane opening a_0 under rated power and the interval of guide vane opening shall be 10% of a_0. The interval of guide vane opening may be lager for sediment water.

6.1.3 $\eta = f(n_{11})$ curve shall be verified repeatedly with clean water and sediment water at least at two guide vane openings. The operating case at one guide vane opening shall be at the optimum efficiency zone.

6.1.4 For model turbine with adjustable runner blades, the interval of blade angle may be set at 5° during the test. The blade angle under optimum efficiency condition, rated load condition and design condition shall be verified repeatedly.

6.2 Cavitation Test

6.2.1 Besides the measurement of critical cavitation coefficient σ_c of clean and sediment water, the position where bubbles occur on the blade and their development process with clean water shall be observed under incipient cavitation, plant cavitation and critical cavitation coefficients. Coverage area, length and other properties of cavity shall be described and photographed. Incipient cavitation with sediment water shall be observed by means of acoustic analysis. Acoustic analysis shall be verified with observed results obtained by means of drum stroboscope under clean water condition.

6.2.2 The characteristic curve of the cavitation test shall take the cavitation coefficient σ as X-coordinate and take the efficiency η, unit power P_{11} and unit discharge Q_{11} as Y-coordinate respectively. The following three curves shall be plotted: curve $\eta = f(\sigma)$ of efficiency versus cavitation coefficient, curve $P_{11} = f(\sigma)$ of unit power versus cavitation coefficient, curve $Q_{11} = f(\sigma)$ of unit discharge versus cavitation coefficient.

6.2.3 The critical cavitaiton coefficient σ_c should be determined by the curve $\eta = f(\sigma)$ together with both two curves of $P_{11} = f(\sigma)$ and $Q_{11} = f(\sigma)$ for reference.

6.2.4 Critical cavitaiton coefficient σ_c may take as σ_0 or σ_1 (see Figure 6.2.4) and it shall be specified clearly in the test program.

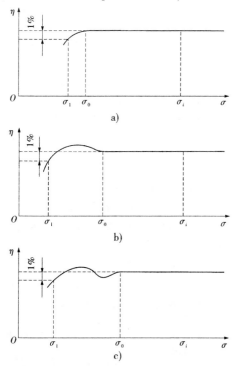

Figure 6.2.4 Common methods for determining σ_c

6.3 Sand Erosion Test

6.3.1 The test for the position of sand erosion and the relative erosion intensity of hydraulic passage parts should be carried out by using easily-damaged coating method. If sand erosion test is performed directly on the flow surface of a model turbine, the test shall be carried out on other spare parts including runner, guide vane, and seal ring, etc. The dimension of hydraulic passage, efficiency and cavitation characteristics of the spare parts shall be consistent with those in the performance test.

6.3.2 If the easily-damaged coating method is used in the sand erosion test, coating should be made with different colored layers. The coating surface shall be smooth and the layer thickness shall be uniform for each layer. The total coating thickness should not exceed 0.5 mm. The physical and chemical properties shall be steady. The coating has good water resistance and it is easy to be cleaned off after a test. Wear resistant capacity of the coating shall meet the requirements of Clause 6.3.3.

6.3.3 It takes at least 30 min to conduct the sand erosion test under every operating condition. If the elapsed test time exceeds 3 h, the sediment shall be replaced every 3 h and the sediment concentration shall meet the requirements of Clause 5.6.4 after every replacement.

6.3.4 Repeated sand erosion tests shall be carried out at least for one operating condition. The permissible deviation of erosion intensity at corresponding position of two tests shall be within ±25%.

6.3.5 Relative erosion intensity at different positions shall be analyzed according to erosion area and erosion depth after an erosion test. And the results shall be the reference for anti-wear protection of prototype turbine and the improvement of hydraulic designs of model turbines.

6.4 Other Tests

6.4.1 Runaway speed tests should be carried out under plant cavitation coefficient with clean water.

6.4.2 The pressure fluctuation with clean water and sediment water shall mainly focuses on the specified points of the draft tube. The amplitude and frequency of pressure fluctuation in the spiral case and on head cover may also be verified. Pressure fluctuation tests shall be carried out under plant cavitation coefficient.

6.4.3 Force characteristic tests should be performed under plant cavitation coefficient. The test is to verify axial hydraulic thrust, total hydraulic pressure and hydraulic torque on guide vane. The maximum torque on blade shall be also verified for model turbines with adjustable runner blades.

7 Verification of Guarantees

7.0.1 Various data shall be provided before test, including hill diagram, curves of $\eta = f(n_{11})$ and $\eta = f(\sigma)$, runaway speed curve and pressure fluctuation curve at different guide vane openings, other curves and data related to verification.

7.0.2 The results of model tests may be taken as guarantees. And based on the results of model tests, the corrected results according to scale effect may be taken as guarantees of prototype turbines. The efficiency correction method to consider scale effect shall meet the requirements of Clause 7.0.4.

7.0.3 Under plant cavitation coefficient and different sediment concentration conditions, efficiency, output power and weighed average efficiency under maximum head, design head, rated head and minimum head shall be verified.

7.0.4 The efficiency of the prototype turbine may be obtained from that of the model turbine and calculated according to Equation (7.0.4-1), Equation (7.0.4-2) and Equation (7.0.4-3).

$$\Delta \eta_{h.P} = \eta_{h.M} + \Delta \eta_h \tag{7.0.4-1}$$

$$\Delta \eta_h = \delta_{ref} \left[\left(\frac{Re_{ref}}{Re_M} \right)^{0.16} - \left(\frac{Re_{ref}}{Re_P} \right)^{0.16} \right] \tag{7.0.4-2}$$

$$\delta_{ref} = \frac{1 - \eta_{h.opt.M}}{\left(\frac{Re_{ref}}{Re_{opt.M}} \right)^{0.16} + \frac{1 - V_{ref}}{V_{ref}}} \tag{7.0.4-3}$$

where $Re_{ref} = 7 \times 10^6$;

$\eta_{h.P}$—calculated hydraulic efficiency of prototype turbine, %;

$\eta_{h.M}$—hydraulic efficiency of model turbine, %;

Re_M—model Reynolds number for each operating condition;

Re_P—prototype Reynolds number for similar operating conditions;

δ_{ref}—relative scalable loss corresponding to Re_{ref};

$\eta_{h.opt.M}$—the optimum hydraulic efficiency of model turbine with each test sediment concentration, %;

$Re_{opt.M}$—Reynolds number corresponding to $\eta_{h.opt.M}$;

V_{ref}—loss distribution coefficient corresponding to Re_{ref}, for turbines with adjustable runner blades, $V_{ref} = 0.8$, and for Francis turbines and fixed blade turbines, $V_{ref} = 0.7$.

For different test sediment concentrations, efficiency conversion from the model to the prototype shall be based on the corresponding $\eta_{h.opt.M}$ and $Re_{opt.M}$.

7.0.5 Verification of cavitation guarantees: the plant cavitation coefficient, critical cavitation coefficient and incipient cavitation coefficient under the specified operating conditions shall be verified. The cavitation picture with clean water shall be provided or the location and pattern of cavitation are described by the simplified diagram under those operating conditions.

7.0.6 For more than two model turbines or runners, sand erosion comparison tests shall be carried out for each model turbine or runner under the same sediment concentration condition. Relative erosion resistant capacity shall be analyzed according to the respective test conditions after the test and the model turbine or runner with better erosion resistant capacity shall be recommended.

7.0.7 Other parameters conversion. The discharge Q_P, power P_P and rotational speed n_P of prototype turbine may be converted from test data of model acceptance tests and calculated according to Equation (7.0.7-1), Equation (7.0.7-2) and Equation (7.0.7-3).

$$Q_P = Q_M \left(\frac{H_{n,P}}{H_{n,M}}\right)^{1/2} \left(\frac{D_P}{D_M}\right)^2 \qquad (7.0.7\text{-}1)$$

$$P_P = P_M \left(\frac{H_{n,P}}{H_{n,M}}\right)^{3/2} \left(\frac{D_P}{D_M}\right)^2 \frac{\eta_{h.P}}{\eta_{h.M}} \qquad (7.0.7\text{-}2)$$

$$n_P = n_M \left(\frac{H_{n,P}}{H_{n,M}}\right)^{1/2} \left(\frac{D_M}{D_P}\right) \qquad (7.0.7\text{-}3)$$

where, the subscripts M and P indicate model and prototype respectively.

Annex A Density of Mercury, Physical Property of Water and Acceleration Due to Gravity

Table A-1 Density of mercury at different temperatures

Temperature (℃)	0	5	10	15	20
Density (kg/m³)	13,596	13,583	13,571	13,559	13,546
Temperature (℃)	25	30	35	40	45
Density (kg/m³)	13,534	13,522	13,509	13,497	13,485

Table A-2 Vapour pressure of water at different temperatures

Temperature (℃)	Vapour pressure (m H₂O)	Temperature (℃)	Vapour pressure (m H₂O)	Temperature (℃)	Vapour pressure (m H₂O)
0	0.062,3	12	0.143,0	24	0.304,9
1	0.067,0	13	0.152,7	25	0.323,8
2	0.071,9	14	0.163,0	26	0.343,7
3	0.077,2	15	0.173,9	27	0.364,7
4	0.082,9	16	0.185,5	28	0.386,7
5	0.088,9	17	0.197,7	29	0.410,0
6	0.095,3	18	0.210,6	30	0.434,4
7	0.102,1	19	0.224,3	31	0.460,1
8	0.109,3	20	0.238,7	32	0.487,1
9	0.117,0	21	0.254,0	33	0.515,5
10	0.125,2	22	0.270,0	34	0.545,4
11	0.133,8	23	0.287,0	35	0.576,7

Table A-3 Density of clean water at different temperatures

Temperature (℃)	Density (kg/m³)	Temperature (℃)	Density (kg/m³)	Temperature (℃)	Density (kg/m³)
0	999.80	12	999.48	24	997.32
1	999.88	13	999.34	25	997.10
2	999.92	14	999.20	26	996.84
3	999.96	15	999.00	27	996.56
4	1,000.00	16	998.88	28	996.30
5	999.98	17	998.72	29	996.00
6	999.94	18	998.54	30	995.70
7	999.90	19	998.36	31	995.36
8	999.84	20	998.20	32	995.00
9	999.78	21	997.96	33	994.64
10	999.70	22	997.74	34	994.26
11	999.60	23	997.54	35	993.90

Table A-4 Dynamic viscosity of clean water at different temperatures

Water temperature (℃)	Dynamic viscosity (10^{-3} N·s/m²)	Water temperature (℃)	Dynamic viscosity (10^{-3} N·s/m²)	Water temperature (℃)	Dynamic viscosity (10^{-3} N·s/m²)
0	1.781	15	1.139	30	0.798
5	1.518	20	1.002	35	0.719
10	1.307	25	0.890	40	0.653

Table A-5 Acceleration due to gravity at different latitudes and elevations

Latitude (°)	Elevation (m)				
	0	1,000	2,000	3,000	4,000
	Acceleration due to gravity (m/s²)				
0	9.780	9.777	9.774	9.771	9.768
10	9.782	9.779	9.776	9.773	9.770
20	9.786	9.783	9.780	9.777	9.774
30	9.793	9.790	9.787	9.784	9.781
40	9.802	9.799	9.795	9.792	9.789
50	9.811	9.808	9.804	9.801	9.798
60	9.819	9.816	9.813	9.810	9.807
70	9.826	9.823	9.820	9.817	9.814

Annex B Acceptance Test Program and Report

B.0.1 The acceptance test program shall mainly include the following items:

    a Description of project.

    b Project program and plan schedule.

    c All necessary documents, including layout patterns and dimensions which possibly affect the hydraulic performance of the turbine shall be provided before acceptance test.

    d The requirements of test rig and test schedule.

    e Test items of the model acceptance test.

    f Check of model turbine and calibration of measurement instruments.

    g Sediment concentration, median grain size, group and operating conditions of acceptance test are specified in each acceptance test.

    h The check and acceptance shall include various performance curves, data, chart, geometric shape and size of hydraulic passage parts.

B.0.2 Acceptance test report shall mainly include the following items:

    a The records, documents and data related to the model acceptance test.

    b Personnel attending the acceptance test.

    c Items of the acceptance test.

    d Specifications of the model test rig and model turbine.

    e Acceptance test procedure and introduction of calibration method of measurement instrument.

    f At least one calculated example in detail to illustrate the testing process from the original data to the final results.

    g Calibration of measurement instruments, uncertainty analysis of parameter measurement related to the acceptance test.

    h Measurement results of geometric size of flow passage of model turbine.

i Various test curves and charts.

j Comparisons between test results and guarantees and technical conclusions.

k Signature of person in charge.

Explanation of Code Terminology

Code terms	Equivalent expression in special situations	Degree of strictness
Shall	Be necessary, require, need, only...permit/allow	Requirement
Shall not	Not allow, not permit	
Should	Recommend, propose	Recommendation
Should not	Not recommend, not propose	
May	Allow, permit	Permission
Need not	Not require, not need	